# Wind Power

## Ed Catherall

**Silver Burdett Company**

# Fun with Science

Electric Power    Hearing          Clocks and Time
Solar Power      Sight              Levers and Ramps
Water Power     Taste and Smell   Magnets
Wind Power      Touch             Wheels

DB

First published in 1981 by Wayland Publishers Limited
49 Lansdowne Place, Hove, East Sussex BN3 1HF, England

© Copyright 1981 Wayland Publishers Limited
Published in the United States by
Silver Burdett Company, Morristown, N.J. 1982 Printing

ISBN 0-382-06628-6
Library of Congress Catalog Card No. 81-86271

Illustrated by Ted Draper
Designed and typeset by DP Press Limited, Sevenoaks, Kent
Printed in Italy by G. Canale & C.S.p.A., Turin

# Contents

## Wind direction

Which way is the wind blowing?

How many different ways can you think of to find out which way the wind is blowing?

Are the clouds moving in the same direction as the wind on the ground?

From which direction does the wind usually blow?

You can find out the wind direction by looking at a weather vane.

How does a windsock tell you the direction of the wind?

How can you tell the strength of the wind by looking at your windsock?

Have you ever seen a windsock at an airfield?

# Measuring the speed of the wind

Glue a length of thread or fine nylon fishing line to a new table tennis ball. Glue the other end of the thread to the centre point of a protractor. Check that the ball swings freely. Tape a small spirit level to the top of the protractor.

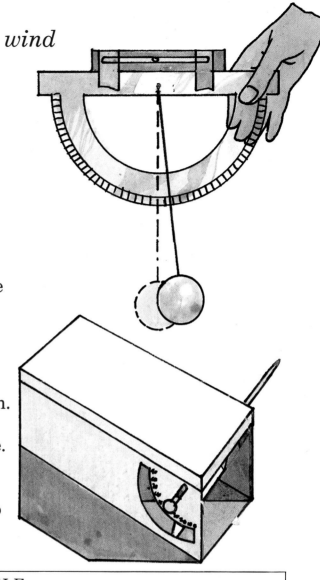

Hold your wind-speed indicator (anemometer) in the wind. Get a friend to read off the angle that the ball is deflected by the wind.

Cut the ends from a shoe box. At one end, make two holes for a knitting needle.
Place the knitting needle in position. See that it spins freely.
Tape a cardboard flap to the needle. Cut a slit so that you can see the position of the flap.
Place your shoe-box anemometer to face the wind. Record the position of the flap.

| ANGLE | | | | | | | | | | | | | | |
|---|---|---|---|---|---|---|---|---|---|---|---|---|---|---|
| 90 | 85 | 80 | 75 | 70 | 65 | 60 | 55 | 50 | 45 | 40 | 35 | 30 | 25 | 20 |
| MILES PER HOUR | | | | | | | | | | | | | | |
| 0 | 5.8 | 8.2 | 10.1 | 11.8 | 13.4 | 14.9 | 16.4 | 18.0 | 19.6 | 21.4 | 23.4 | 25.8 | 28.7 | 32.5 |
| KILOMETRES PER HOUR | | | | | | | | | | | | | | |
| 0 | 9.3 | 13.2 | 16.3 | 19.0 | 21.6 | 24.0 | 26.4 | 29.0 | 31.5 | 34.4 | 37.6 | 41.5 | 46.2 | 52.3 |

# The Beaufort Scale

In 1805, Sir Francis Beaufort devised a scale for wind speeds.

| N | Description | Wind speed | | Common signs for recognition |
|---|---|---|---|---|
| | | mph | km/h | |
| 0 | Calm | 0–1 | 0–1.6 | Smoke rises vertically. |
| 1 | Light air | 2–3 | 3.2–4.8 | Smoke slowly drifts. |
| 2 | Light breeze | 4–7 | 6.4–11.3 | Wind felt on face; leaves just move. |
| 3 | Gentle breeze | 8–11 | 12.8–17.7 | Flags flap; leaves move continuously. |
| 4 | Moderate breeze | 12–16 | 19.3–25.7 | Paper blows; dust raised; small branches move. |
| 5 | Fresh breeze | 17–21 | 27.4–33.8 | Small trees in leaf sway. |
| 6 | Strong breeze | 22–27 | 35.4–43.5 | Branches on trees move. |
| 7 | Moderate gale | 28–33 | 45.1–53.1 | Whole trees sway. |
| 8 | Fresh gale | 34–40 | 54.7–64.4 | Twigs and small branches break off. Gale warnings on radio. |
| 9 | Strong gale | 41–48 | 66.0–77.3 | Large branches break off. Slight damage to property. |
| 10 | Whole gale | 49–56 | 78.9–90.2 | Trees uprooted; major damage. |
| 11 | Storm | 57–65 | 91.8–104.7 | Danger – take shelter. |
| 12 | Hurricane | 66+ | 106.3+ | Usually at sea or coastal areas. Disaster conditions. |

Look in the news media for reports of storms, hurricanes, typhoons and freak wind conditions.
Where did these winds occur?
What was the maximum wind speed?
What time of year was it?
How much damage was done?

# The wind changes things

Have you seen trees growing away from the wind?
The trees are growing away from the usual wind direction. They are leaning with the 'prevailing wind'.

Put some fine, dry sand into a large tray. Using a straw to direct the wind, blow the sand very gently. Notice how the sand moves and makes patterns.
What is the effect of wind on a sandy beach?

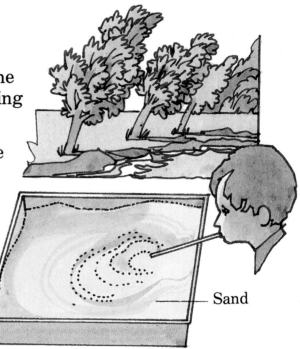

Sand

How do sand dunes form?
What is the shape of a sand dune?
Do sand dunes move?

Sand dune

The wind causes sand storms and dust storms.

Snow fence

How do snow fences work?
Why are they placed in a particular position?

# Chapter 2 The wind acts as a brake

## The wind acts as a brake

Go out on a windy day and feel the
force of the wind.
Try running *into* the wind.
Feel how the wind slows you down.
The wind is acting as a brake.

Try running *with* the wind. Be
careful that the wind does not blow
you over. What happens when you
run at angles to the wind?

Try carrying a large sheet of strong cardboard on a
windy day. Try moving against the wind.
Now turn and move with the wind. Hang onto the
card or the wind will blow it out of your hands.
Can you move at different angles to the wind?
How do you have to move the cardboard?

## Making an air brake

Cut four equal slits at right angles on one
end of a cotton reel.
Place a square of thin card in each slot.
Put a knitting needle through the
centre of the cotton reel.
Hold the knitting needle and blow
on the card squares. See that the
cotton reel spins freely.
Remove the card squares.
Attach a hook made from a paper
clip to the far end of the thread.
Wind up the hook.
Place a washer on the hook as a weight.

Hold the cotton reel and watch the
hook unwind. Wind it up again.
Now time how long it takes to unwind.
How long does it take to unwind
with two, three and four washers?
Record the times.

Now replace the cards in the slots.
Place one washer on the hook.
Does the hook unwind?
Time how long it takes to unwind
with two, three and four washers.
Notice how the air acts as a brake.

What happens if you use smaller squares of card?
What happens if you use larger squares of card?
If you blow on the card, can you use the wind to
wind up the washers?
How many washers can you lift by blowing?

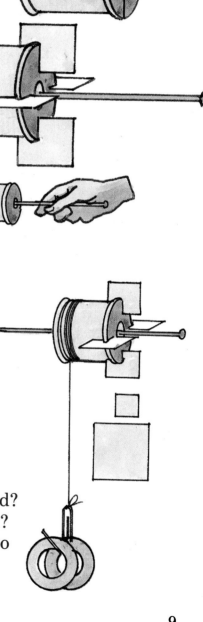

9

## Parachutes

Cut a square from a sheet of thin plastic (polythene).
Cut four equal lengths of cotton and tie one to each corner of the plastic square.
Make a hook from a paper clip and tie it to a single thread. Connect this thread to all four lengths of cotton thread.

Place a washer on the hook and release it from a known height.
Record the time taken for your parachute to reach the ground.
Measure how far the parachute has drifted from the vertical.
Alter the height from which your parachute is released. Record the time.
Add more washers to the hook.
Record the time the parachute takes to fall.
Cut a small hole in the centre of the parachute. Record the falling time.

Make similar parachutes from different cloth materials. Notice how the rate of fall depends on the way the cloth is woven.

Blow some dandelion seeds in the wind.
How are they like parachutes?

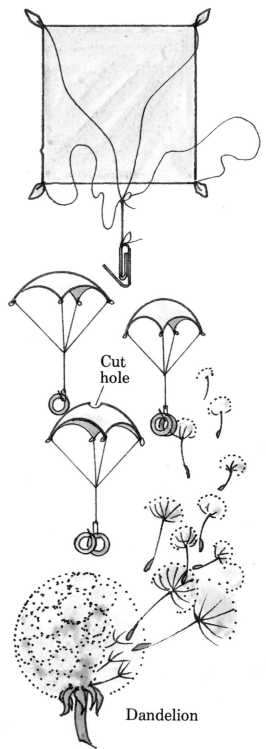

Cut hole

Dandelion

10

# Moving through the air

How far can you throw a ball? The distance that you can throw depends on many factors.

It depends on how strong you are and how you get the power in your arms transferred to the ball. Practising throwing helps you develop a good throwing style.

It depends on how fast you can throw a ball. The faster the ball leaves your hand, the further it will go.

It depends on the angle that you throw the ball at. If you throw the ball flat, it will fall quickly. If you throw it too high, you are using energy to make the ball go up rather than along.

It depends on the air. If the wind is blowing from behind the ball, it will go further. If you throw the ball into the wind, then the wind will slow the ball down and it will fall quickly to the ground.

It depends on the size of the ball. The larger the ball, the more the wind will slow it down.

Go outside and practise throwing a ball. See how the wind affects your throw. What happens if you spin the ball?

# Making paper darts

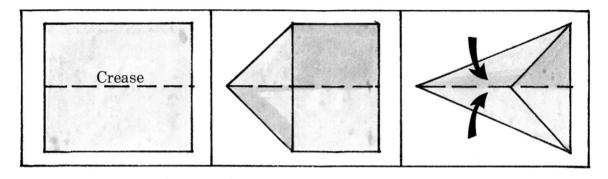

Cut an oblong of paper. Fold the sides to
make a crease down the middle.
Fold the two top corners down into the
middle so that they just touch.
Fold the two triangles inwards again
so that they just touch in the mid crease.
Raise the sides by folding the first crease
again.

Now fold the wings down at right angles to
the body, so that the centre crease is a
keel to hold your dart by.

Launch your dart and see how it flies.

Cut two slits at the back of each wing of
the dart. Lift up the flaps.
How does this affect the flight of the
dart?
These flaps act as air brakes or ailerons
and make the dart roll.

Try flying your dart with one aileron up
and the other down.

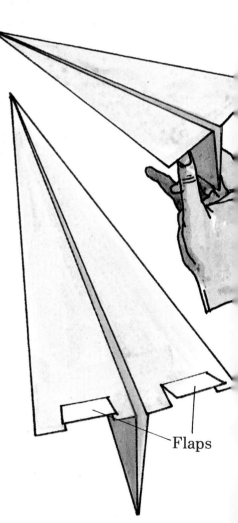

Flaps

## Making a spinner

Cut a square from a sheet of thin card.
Divide one side into thirds and make two
straight cuts.
Lift the two outer strips to form wings.
Place a paper clip on the bottom edge of
the centre section.
Release your spinner from a known height.
Record how long it takes to fall.

Alter the wings so that the spinner
increases the number of turns. Notice how
this slows the rate of fall.

Can you make your spinner spin the
other way?

What happens if you put two paper clips
on your spinner?

Badminton shuttles also use air resistance
or air acting as a brake.
Compare the flight of a plastic shuttle with
that of a feather shuttle.

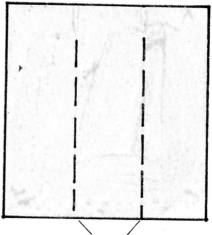

Cuts

Paper clip

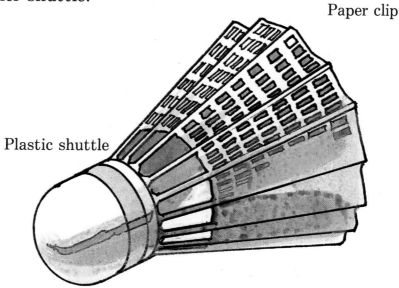

Plastic shuttle

## Streamlining

Fix a short candle into a saucer or tin lid. Light the candle.
Blow gently and see the effect of the wind on the candle flame.

Place a large book or card in front of the candle. Blow again and see how the book shields the candle from the wind.
Move the book (or card) further from the candle. Take away the book (or card) and replace it with a cylindrical can.
Now blow. What happens? Move the can closer or further from the candle. What happens?

There is a lot of wind around tall, square buildings. Go outside and blow bubbles into the wind near a tall building. Notice how the bubbles travel around and over buildings.

Notice how sports cars are streamlined to move through the air.

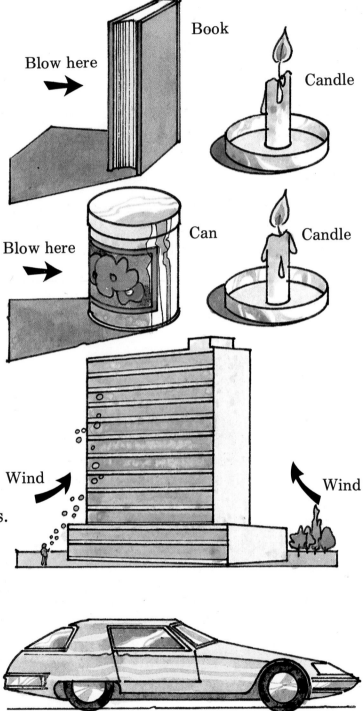

14

# Chapter 3  Using the wind

## Making a mobile

Draw sea creatures of different sizes on card. You could draw a shark, small fish, sea horses and sea serpent.
Cut each one out carefully and paint your sea creatures on both sides.

Make a hole at the top of each creature and tie each one to a length of cotton.
Tie the two smallest to a length of fine cane or strong wire.
Find the balance point. Tie another length of fine cotton to the balance point.
Tie the whole system to the end of another length of cane or wire.
Tie another, larger creature at the other end of the length of cane or wire.
Now find the balance point of the new complete system and tie a length of cotton to this balance point. You may need to weight the larger creature with paper clips.
If you need to anchor the cotton, use spots of instant glue.

Hang your mobile in a gentle wind and watch your sea creatures move.
You can make larger mobiles and, by using small bells, make a sound mobile.

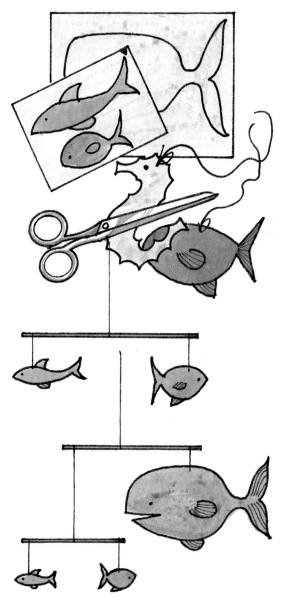

## Making a windmill

Cut a square of card.
Draw pencil lines from corner to corner.
Make small pinholes in the four corners
and one where the pencil lines cross.
Cut half way along each pencil line
starting from the outer edge.
Bend over the angle pieces so that
the four holes are in line with the
centre hole. Put a pin through all of
the angle holes and then into the
middle hole.
You have made a windmill.
Put a plastic bead onto the pin
behind your windmill.
Push the pin firmly into a strong
stick.

Blow the windmill from the front.
Which way does the windmill turn?

Move your windmill slowly through
the air. Which way does your
windmill turn?

What happens if you blow gently
and then strongly on your windmill?

Can you make a windmill that spins
the other way? If so, fix it on the
stick beneath your first windmill.

Have you ever seen a full-sized windmill?
Where did you see it and what was it used for?

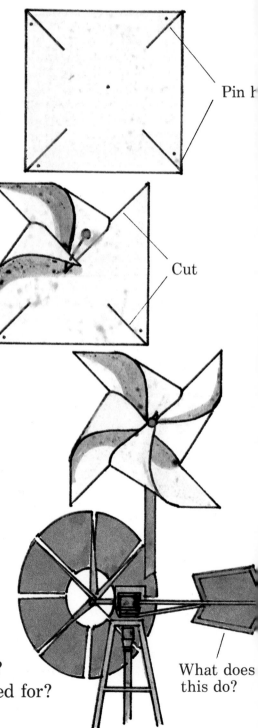

Pin h

Cut

What does
this do?

16

# Kites

**Danger**: Keep your kites away from all overhead wires.

One of the best ways to understand wind is to fly a kite.
There are many kinds of kites.
The controllable kite is one of the easiest to fly.
When flying your kite, ask yourself some questions:

How do I get the kite to climb?

How do I make the kite swoop?

Will the kite fly if there is no wind?

What weight will the kite carry?

Where must I fix the weight?

What happens if I alter the tail?

How many different kinds of kites have you seen?

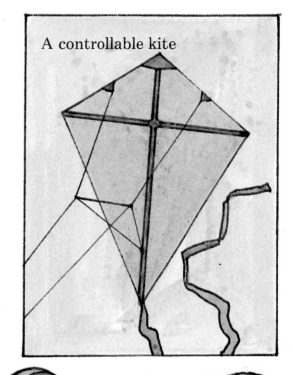

A controllable kite

17

## Sails and yachts

Cut out a boat shape from a thick piece of balsa wood.

Ask an adult to help you make a keel from a can lid. Use an old pair of scissors to cut the keel.

Push the keel into the middle of your boat.

Use a length of thin dowel for the mast. Push the mast into the top deck of the boat, right in the centre, but in front of the keel.

Cut out a triangle of thin plastic for the sail. Tape the sail to the mast.

Tie a short length of cotton to the end of the sail and wind this cotton around a tack.

Now launch your boat. If it tends to turn over you can either make the mast and sails smaller or add paper clips to the end of the keel.

Does your boat sail downwind?
How can you make your boat sail at an angle to the wind?
Will your boat sail into the wind?

If you take out the keel and fit four wheels to your boat, you will make a land yacht.

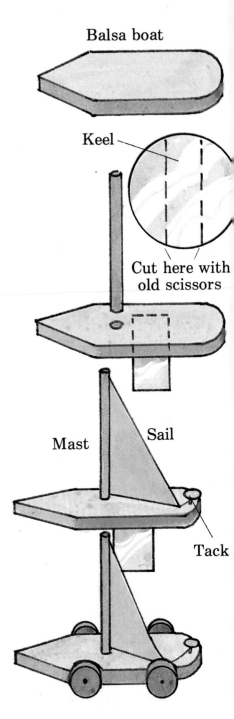

Balsa boat

Keel

Cut here with old scissors

Mast

Sail

Tack

18

## Some experiments with wind

**Experiment 1**
Cut an oblong, 20 cm by 6 cm, from
stiff paper.
Fold down 3 cm from each side to make
a bridge.
Place your paper bridge on a flat surface
Blow steadily under the bridge.
What happens to the top of the bridge?
What happens to the sides of the bridge?

**Experiment 2**
Cut a 10-cm square of card.
Draw pencil lines from corner to corner.
Push a pin into the card where the lines
cross. Then push the pin into the hole
in a cotton reel.
Hold the card tightly against the cotton
reel while you blow steadily up the
centre of the cotton reel from the other
end.
While blowing, take your hand away.
Can you get the card to stay up?

**Experiment 3**
Punch a hole in the base of two equal
size cans.
Thread string through each hole and
knot it.
Suspend the cans at the same height
but only 3 cm apart. Blow gently
between the cans.
Which way do the cans move?

Blow here

Blow
here

19

## *Wings*

Hold a sheet of paper close to your mouth.
Hold it so that the paper hangs down in a curve.
Blow gently across the top of the paper.
What happens to the paper?
Take a sheet of paper and fold it in half.
Crease the fold firmly.
Curve the top sheet and glue it in place.
You have made a wing.

Blow here

Make a small hole in the top and
bottom of your wing. Cut a short
length from a straw to prop up
your wing and stop it collapsing.
Thread cotton through the holes
and the straw.
Hold the cotton very firmly in
both hands.
Pull the cotton vertical. Now,
while pulling the cotton, blow
from the crease.
What happens to the wing?
Holding the cotton vertical, move
the wing rapidly forward so that
the wind travels in the same
direction as you blew.
What happens to the wing?
What happens if you blow from
the other direction?
What happens if you turn the
wing upside down?

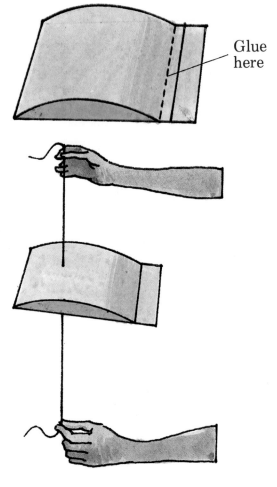

Glue here

## How wings lift

The distance over the top of an aerofoil is longer than under it. The air that goes over the top has further to go than the air underneath. The top air is then spread out more, so it has less pressure than the air underneath. The air pressure underneath pushes the wing into the area of reduced pressure.

Collect bird feathers. You will find that there are basically three kinds of feathers: down feathers to keep the bird warm; tail feathers that act as a brake; and flight feathers that give the bird 'lift' so that it can fly. Notice that the tail feather has a mid rib that runs down the centre, while the mid rib of the flight feather is not central.

Push a pin into the quill of a flight feather. Hold the pin and move the wing feather in the air. Notice how the wing always keeps the same position.
Now try a tail feather. What happens?

A section
of an aerofoil

Longer distance

Reduced air pressure

Direction
of lift

Shorter distance

Tail feather

Flight feather

## Feathers and wings

Make four holes at right angles to each other near the top of a cork. Put a tail feather into each hole. Weight your cork with a tack. Drop your toy and watch it float down like a helicopter. Make four similar holes in another cork. Put a wing feather into each hole. Make sure each feather is facing the same way. Weight this cork with a tack. Drop this toy. Why does it spin?

Look closely at the tail and wing feathers.
Look carefully at the wing of a dead bird. See how the wing is an aerofoil.
Note the position of the main flight feathers.
Watch birds flying. Notice when they flap their wings and when they glide.

A bird wing

Watch sycamore or maple seeds fall from a tree. Why do they spin?

A maple seed

# Gliders and helicopters

Gliders have long wings. Each wing is an aerofoil.
The glider is catapulted or towed into the air. The pilot looks for upward air currents so that he can climb higher. Given the right wind conditions, gliders can stay up for days.

Have you ever seen a hang glider? Here, the delta-wing sail acts as an aerofoil.
Try to find out how hang gliders are flown and how they are controlled in flight.

When a yacht sails close to the wind, the sail acts as an aerofoil. This sucks the boat along. The keel stops the boat going sideways.

Look at the rotor blades of a helicopter. Each is an aerofoil. When they turn fast, the helicopter lifts. It is working like your feather toy (page 22).

Racing cars use upside-down aerofoils to press the back wheels onto the track at high speed.

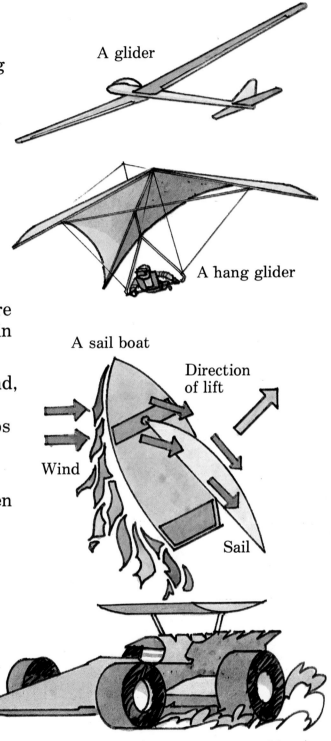

A glider

A hang glider

A sail boat

Direction of lift

Wind

Sail

A racing car

## Propellers

Buy a plastic propeller.
Cut two equal lengths of
strong wire.
Push each length of wire
through the ends of a 20-cm
long strip of balsa wood.
Cut the end from a ballpoint
pen.
Push a short length of wire
through the propeller, a
bead, the outer case of the
ballpoint pen and then make
a hook.

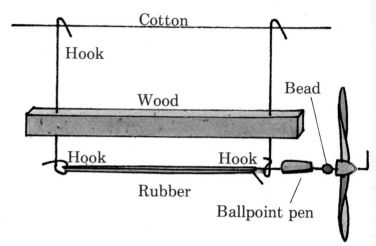

Connect the wire through the wood to the propeller
wire, just behind the pen.
Connect rubber bands to the hook and the back wire.
Connect your machine to a length of cotton that is taut
and horizontal by bending over the wire as a loop.
Wind up your propeller. Release it. Which way did the machine go?
Wind your propeller the
other way.
What happens?

You can turn the machine you
have made into a land
machine.

Propellers work like fans.
They make a wind, pushing
the air backwards (see page 16).

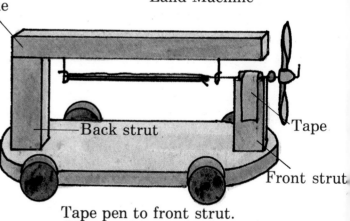

Land Machine

Tape pen to front strut.
Glue machine to back strut.

24

## Jet propulsion

Thread cotton through a plastic straw. Fasten each end of the cotton to supports.
Make sure the cotton is taut and horizontal.
Put two short lengths of straw in the neck of a balloon.
Hold the straws in place with a rubber band.
Blow up the balloon and twist the neck to keep the air in.
Fix the balloon to the straw with sticky tape.
Untwist the neck of the balloon and let it go.
How far did your balloon travel?
Blow up your balloon with different amounts of air.
How does this alter the distance your balloon will travel?
Change the number of straws in the nozzle. What happens?
Will your balloon travel uphill?
Will your balloon carry weights?

Make a balsa wood boat and a keel (see page 18).
Make a hole near the stern of your boat.
Stick the neck of a balloon through the hole. Blow up the balloon. Twist the neck of the balloon. Fix the keel to the boat.
Place the boat in the water and untwist the neck of the balloon.

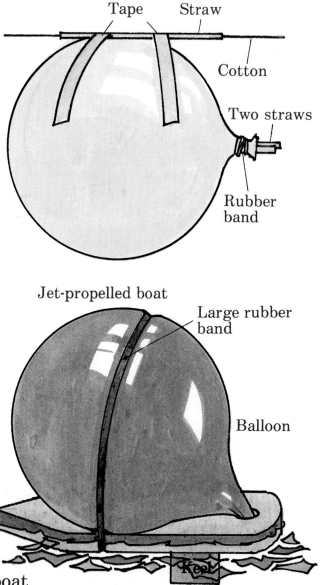

Tape   Straw

Cotton

Two straws

Rubber band

Jet-propelled boat

Large rubber band

Balloon

Keel

## Hovercraft

Cut out a 6-cm hole from the centre of a polystyrene ceiling tile. Put the tile on a flat, smooth surface. Blow down into the hole. Can you get your tile to lift?

Use a hairdryer to blow into the hole in the tile.

To get the tile moving give it a slight push.

Cut the corners from the tile to make it boat shaped.

Glue the corners onto the top of the tile to act as rudders.

Real hovercraft ride on a cushion of air made by large fans.

Try to find out where hovercraft are used.

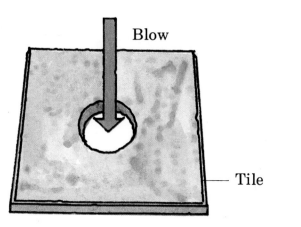

Blow

Tile

Blow

Corner

# Hot-air balloons

Take three large sheets of tissue paper. Overlap them, and glue the overlaps until you have made a large sheet.
Do this six times until you have made six large sheets.
Wait for the glue to dry. Check that there are no holes at the overlap.

Put all six sheets in a pile and carefully cut out the shape of a balloon panel.

Carefully glue the six panels together to make your balloon. You will have to glue very carefully to see that there are no leaks between the panels.

Fold another sheet in four to make a thick strip. Glue this strip at the bottom of the balloon to make a collar. You might have to strengthen the top of the balloon with a tissue-paper disc.

Use air from an electric convector fire or large hairdryer to inflate your balloon. Hold your balloon until you feel the lift, then release it. The larger the balloon the further it goes.

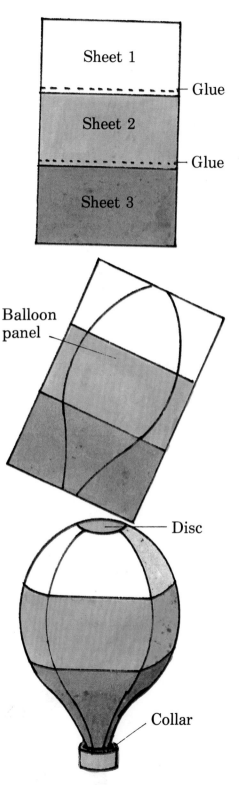

Sheet 1

Glue

Sheet 2

Glue

Sheet 3

Balloon panel

Disc

Collar

## The wind dries up water

On a windy day, take two
handkerchiefs of the same size. Soak
them both in water.
Hang one handkerchief in a windy place.
Hang the other in a sheltered place
where the wind cannot reach it.
Feel the handkerchiefs every fifteen
minutes.
How long does the handkerchief in
the windy place take to dry?
How long does the handkerchief in
the sheltered area take to dry?
Where has the water gone?

Dissolve some salt in water.
Pour the salt water into a saucer.
Leave the saucer in a windy place.
How long does it take for all
the water to evaporate?
Can you see the salt left in
the bottom of your saucer?

The wind is drying up the
water all over the world. A
lot of this water in the air
eventually forms clouds.
These clouds then give us
mist, rain, snow or hail,
which are all forms of water
that will dry up again.

## The wind on your skin

When you sunbathe, the wind dries your skin. To prevent this, you have tiny oil glands in the skin to lubricate the hairs. This makes the skin oily and cuts down evaporation.

You can also rub your skin with moisturizing creams to replace the water lost.

Normally, the hairs on your skin lie flat. When we are in cold air, hair muscles contract and each hair is pulled up. This gives our skin the 'goose bump' look and thickens our covering of body hair. Birds fluff up their feathers in cold weather.

In hot weather, we wear fewer clothes so that the wind can cool our bodies.

When we are feeling hot, blood is pumped near the surface of our skin so that the air can cool our blood. That is why we look red when we are hot.

In hot weather, sweat glands in the skin produce more sweat.

This sweat is evaporated by the wind and cools us.

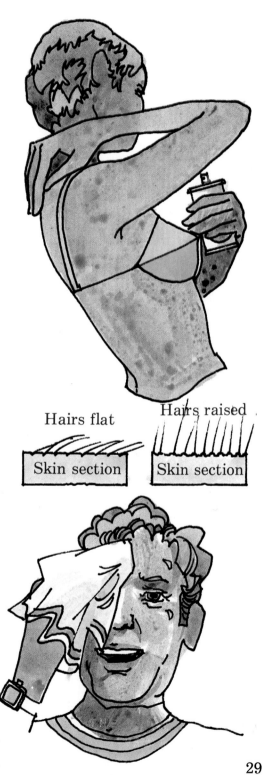

Hairs flat

Skin section

Hairs raised

Skin section

## Cooling

Find two identical flat pans.
Put the same amount of water in each pan.
Put a thermometer into each pan and check that the water temperature is the same.
Make sure that the thermometers are covered with water.
Stand each pan on a pile of newspapers. These act as insulators.
Put one pan in a windy place. Put the other pan in a sheltered place.
Make sure both are in the shade.
You can use a fan to create a wind.
Read the temperature of the water every five minutes.
Which pan has the cooler water?

Take two identical bottles.
Fill them both with the same amount of water.
Wrap each bottle in the same kind of cloth.
Soak the cloth around one bottle, leave the other dry.
Stand both bottles on newspapers in a windy place. Check the temperatures of the water in the bottles.
How much cooler is the water in the soaked-cloth bottle?

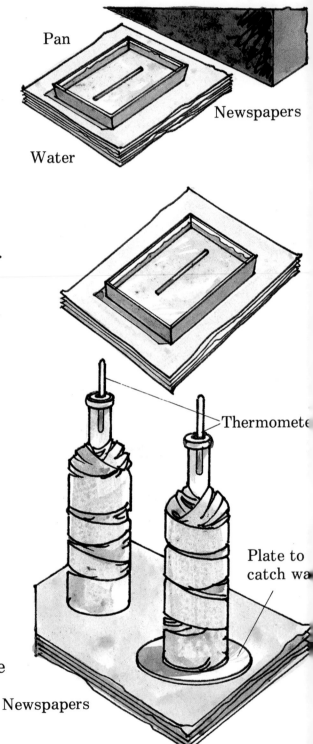

Pan

Newspapers

Water

Thermomete

Plate to catch wa

Newspapers

# Fans

Cut some long, very thin strips of tissue paper. Tape the ends of these paper strips to a stick.

Blow gently on the strips. Do they move?

Use your draught detector to check the draughts in your room. Investigate the doors and windows.

In hot weather, we use fans to create a draught for air conditioning.
Switch on an air-conditioning fan.
Hold your draught detector in front of the fan. Which way is the air blowing?
Hold your draught detector behind the fan.
Which way is the air blowing?

We use fans to blow air in both directions. Look at a vacuum cleaner.
Find out which way the fan makes the air move.
How many machines can you think of that use a fan?

Fan

Vacuum cleaner

A small snow blower

# *Pumping air*

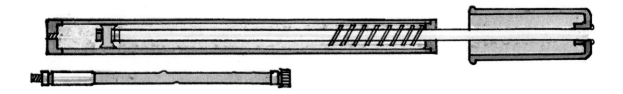

Use a bicycle pump to pump air. What do you have to do to get the air to come out of the nozzle?
Put your finger over the nozzle.
Feel the air coming out as you pump.
Notice how the pump gets warm if you pump fast.
Try to stop the air from coming out of the nozzle as you pump.
What does it feel like?

Notice how the pump is acting like a brake.

Use your pump to put air into a bicycle tyre. What do you have to do?

In 1888, Mr Dunlop, a Scottish veterinary surgeon, invented the first air-filled, or 'pneumatic' tyre.

What do tyres do?
Why should you always keep your tyres pumped up?
What does the tyre valve do?
How does it work?

Carefully take your bicycle pump apart. How does it work?